MW01265596

Perfection Learning®

Cover and inside design: *Michelle J. Glass*

If a species becomes extinct,
its world will never come into being again.
It will vanish forever like an exploding star:
And for this, we hold direct responsibility.

~ David Day

Dedication

For those who dedicate themselves to making "whoop dreams" come true. And to my mother, Mary Goodwin, who was the wind beneath my wings, and my daughter, Margaret, who is sure to soar.

About the Author

Jane Duden taught elementary school in Minnesota and Germany before becoming a freelance writer. She is the author of more than 30 nonfiction books for children. Her favorite topics are animals, science, and the environment.

Ms. Duden's research adventures have taken her as far as Antarctica and as high as the sky in the ultralight airplanes you'll read about in this book. She was privileged to work with whooping crane experts and to have a front-row seat for the Whooping Crane Reintroduction project.

At home in Minneapolis, Ms. Duden likes cooking, swimming, biking, and playing with pets—her own and everyone else's. Best of all, she likes sharing adventures with her now-grown daughter.

Acknowledgements

Special thanks to the Whooping Crane Eastern Partnership, and especially to Heather Ray of Operation Migration

Credits: Operation Migration pp. 1, 3, 4, 5, 6, 7, 11, 12, 13, 14 (left), 15, 17, 18 (bottom), 19, 21, 23, 25, 27, 36, 47, 50–51, 61 (bottom), 62–63, 68 (bottom & top left); International Crane Foundation pp. 16, 52, 56, 61 (top), 67, 68 (top right); Photos.com all sidebars, pp. 10, 31, 32, 34, 35, 38, 39, 40, 42, 54, 55, 57; ArtToday (arttoday.com) behind chapter headings, pp. 9 (left), 14 (right); CORBIS Royalty-Free p. 30; PhotoDisc cloud background pp. 28–49; MapArt™ pp. 26, 44, 58; Photo courtesy of Jane Duden p. 9; Randy Messer p. 18 (top), 20, 64–66 (background)

Printed in the United States of America. For information, contact

Perfection Learning® Corporation
1000 North Second Avenue,
P.O. Box 500, Logan, Iowa 51546-0500.
Tel: 1-800-831-4190 • Fax: 1-800-543-2745
perfectionlearning.com

Paperback ISBN 0-7891-6056-0
Cover Craft® ISBN 0-7569-1387-x

1 2 3 4 5 6 PP 08 07 06 05 04 03

Table of Contents

Introduction

It's quiet in the marsh at Necedah (Nuh SEE duh) National Wildlife **Refuge** (NWR). But not for long. The buzz of an engine breaks the morning peace. The sound gets louder. A strange machine flies into view. It looks like a motorcycle with a wing on top. The pilot is disguised in a baggy white costume.

The tiny plane lands on a grassy strip by the **marsh**. Two more people in white costumes open the gate to a nearby pen. Out steps a tall bird with long legs. Its head and neck are the color of rust. Its body is a spotted mix of white and tan. The bird is a young whooping crane. Four more young cranes follow. They flap their wings, ready to go.

The flying machine speeds down the grassy strip. The birds chase after it. Their strides get longer. They pump their wings. The tiny plane climbs as the graceful whoopers line up off its wing.

These birds are rare and special. They are part of a bold experiment. Soon they will follow that plane on a historic journey.

Wild, migrating whooping cranes vanished from this part of North America over a century ago. Can this tiny plane and dedicated people bring them back?

Whooping Crane

The whooping crane is the rarest and most endangered of the world's 15 crane species. It is North America's tallest bird. The whooping crane got its name because of its loud call. It's like a bugle call that can be heard three miles away. Whoopers can live 25 to 50 years—or even more.

chapter one

Helping Cranes at Risk

From Rare to Vanishing

Few people have ever seen wild whooping cranes in the sky. Whoopers have always been rare. They are the world's most **endangered** crane species. Just 15 whooping cranes were alive in 1941. In 2001, just 260 whooping cranes lived in the wild. A main flock of 173 whoopers **migrates** between Canada and Texas, as whooping cranes have done for **eons**. A small flock of 75 lives year-round in central Florida. Both groups face many risks—**predators**, disease, oil spills, and storms. Human choices about land use could affect them. Even power lines are risks. These wires kill more cranes during migration than any other dangers.

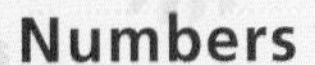

Numbers

Each year the number of whooping cranes changes. Some cranes die and some are born. Scientists hope that the number of cranes born will continue to be more than those that die.

Some additional wild whooping cranes are kept in **captivity** in a few places in North America. Still, no one wants to lose the wild whoopers. With so many of them in a single flock, just one disaster could wipe them out.

A Team to Save Whooping Cranes

Canada and the United States are working together to save whooping cranes. Two crane experts lead the Whooping Crane Recovery Team. Biologist Tom Stehn says, "We need species to survive that have been there since the Ice Age. To keep them alive in captivity—that's just not enough. The Team's aim is to help this endangered species recover so it won't become extinct."

The Team wants to bring whooping cranes back to where they lived long ago. If it works, the species will have a better chance of survival. To carry out the plan, the team formed the Whooping Crane Eastern Partnership (WCEP) in 1999. WCEP (WEE sep) partners are from the United States and Canada. Some are government groups. Some are private groups. They joined forces for one goal—to start a new flock of migrating whooping cranes in eastern North America. The new flock will spend summers in Wisconsin and winters in Florida, as they did long ago. But how will the new flock find Florida when they leave Wisconsin? Who can lead the way?

The Whooping Crane Eastern Partnership Members

- Operation Migration (OM)
- U.S.G.S. Patuxent Wildlife Research Center
- National Wildlife Health Center
- International Whooping Crane Recovery Team
- U.S. Fish and Wildlife Service (USFWS)
- International Crane Foundation (ICF)
- National Fish and Wildlife Foundation
- Wisconsin Department of Natural Resources
- Natural Resources Foundation of Wisconsin
- Canadian Wildlife Service (CWS)

The CWS isn't a founding member, but it has a representative with the WCEP.

Many states and conservation groups help WCEP. Many citizens lend support too.

A Bold Plan: Human-Led Migration

With no wild whooping cranes to lead the way, humans must do the job. Joe Duff and Bill Lishman know how. These two Canadians are **pioneers** in bird migration. It was Bill and Joe's idea to teach orphaned geese to migrate behind an ultralight plane. The 1996 movie *Fly Away Home* was based on their experiment. It showed that the tiny plane could lead young geese on a migration route and the birds would return in the spring on their own.

Bill and Joe started Operation Migration (OM) to research aircraft-led migration. OM pilots worked with geese and sandhill cranes for ten years. They believed they could now lead endangered whooping cranes to a safer future. WCEP partner Operation Migration would teach young whoopers a new migration route.

Ultralight Planes

Ultralight planes can fly at low altitudes. They can fly slow enough to lead birds. The tiny planes are also called *trikes*. At 350 pounds, they are very light. The pilot seat alone on a jet weighs the same as an ultralight!

Sandhill cranes

In 2000, they tested the plan. They used nonendangered sandhill cranes for the test run. Deke Clark joined OM pilots Joe and Bill. They led 11 sandhill cranes from Wisconsin to Florida with ultralight planes. The birds migrated back on their own in spring. They landed right on the grass airstrip where they had taken off six months before!

Migration

Cranes must learn their migration route. They don't know the route unless they are taught. Adult cranes teach their chicks the way south in fall. One time is all it takes. The young will come back on their own in spring. They will head south each fall and north each spring for the rest of their lives.

Thanks to the success with sandhill cranes, Canada and the United States now backed the bold experiment. Twenty states and two provinces approved the plan. The world would soon see the first human-led migration of an endangered species. Would the more fragile and rare whooping cranes let a plane lead the way?

Since the proposed migration is an experiment with an endangered species, it could not start without permits and some new laws. A new rule was added to the Endangered Species Act. It covers all whooping cranes in the new eastern flock. Some of the birds could be lost during the project. So the law protects landowners if one of these whoopers would die on their land. But it is still a crime to kill any whooping crane on purpose.

chapter two

The Plan Gets Off the Ground

Hatching the Flock

The chicks for the WCEP project hatch at Patuxent (Pah TUHK sent) Wildlife Research Center. The center is in Maryland. Its marshes are a good crane **habitat**. They are far from city noise and buildings. Here captive whooping cranes lay eggs. Most chicks for the new eastern flock hatch from eggs laid here. Some eggs are sent from a second **captive-breeding** center, the International Crane Foundation in Wisconsin.

While still in their shells, the chicks hear certain recorded sounds. They hear the purring **contact call** made by adult cranes to their chicks. They also hear an ultralight engine. It helps if the chicks get used to this sound early. Then they won't be afraid of the plane that will teach them to migrate.

Eleven precious eggs hatch between May 7 and May 24, 2001. Their names are the numbers of their hatching order. Biologists in costumes raise the chicks. The birds never see or hear people who are not in costume. The goal is to make sure these cranes learn to avoid humans. Then they will have a better chance of surviving in the wild.

All crane caretakers obey strict rules. They must always wear the white costume around the chicks. They can't talk, burp, laugh, or sneeze. They can't do human actions like carrying things. The birds must never see humans, vehicles, or equipment except for the ultralight. The set of rules for raising and training chicks is called a *protocol.* No one is allowed to "break training."

Protocol

Protocol is the strict set of rules by which the young cranes are raised and trained. The rules include silence around the cranes, dressing in the white suits, and no direct contact. These rules must be followed so **imprinting** with humans doesn't occur.

Ground School

The chicks get their first lessons at Patuxent. Ground school lessons teach them to accept the trike. Dan Sprague is lead trainer for this part of the project. Dan stays with the birds all summer and fall. He stays with the chicks until their migration south is complete.

The trainers make a helper and name it Robo-Crane. It's an extra long puppet. A small **vocalizer** in the puppet plays the contact call of an adult crane. This purring sound means, "It's okay. Come with me."

Chicks think Robo-Crane is their parent. Robo-Crane and other puppets show the chicks how to eat. Like crane parents, the puppets break up any fights.

The trike has a loudspeaker. The pilot pushes buttons to play the crane calls. The contact call is used most. This call makes the chicks want to follow the trike.

A big round pen helps in early lessons. It keeps the chicks safe from the trike while they learn to follow it. The chicks start this training at a few days of age. Dan can work Robo-Crane while he drives the trike on the ground. Dan taxis around the pen. He holds the puppet head out so the chicks can see it. The chicks know the sound of the plane. They follow close beside the puppet head, which plays the contact call. A trigger lets Dan drop **mealworms** from the puppet head.

Mealworms are the chicks' reward from their "parent" for following the trike. They learn fast!

Off to Wisconsin

By early July, the young chicks are almost ready to fly. It's time to take them to Wisconsin to begin their flight training. On July 10, 2001, the ten chicks make their first "migration." Each is put in its own travel crate, and they are all loaded on a private plane. For the first time in more than 100 years, wild whooping cranes arrive in Wisconsin. The staff at Necedah NWR silently welcomes the new arrivals. From now on, the marshes here will be the new eastern flock's summer home.

Eleven to Ten

Female chick 8 died before the flight to Necedah NWR.

Home Sweet Home

It is important that the chicks get to their future nesting grounds in Wisconsin before they learn to fly. Birds never forget the place where they **fledge**. They never forget the first place they can explore from the air. They return there year after year. The site of their fledging will later become their nesting grounds. They must get to know there's no place like home. Their survival depends on a good summer home for nesting and raising the next generation.

Flight School

The ten birds spend the next three months learning to fly with the ultralight. They live in two groups called *cohorts*. Two small groups are easier to train than one large group. The cohort of five older chicks is at one training site. The cohort of five younger chicks is at another site a mile away. Each site has a big pen with a top net for safety at night.

Soon the cranes and planes begin daily training. It's mid-July, and they must be ready to fly to Florida in late October.

Pilots Joe Duff and Deke Clark wake before sunrise each day. They check the weather. They hope the leaves on the trees are still. That means calm air, perfect for training the birds.

First, the older chicks are led from their pen to a grassy strip where the trike is parked. The pilots in crane costume slowly taxi the plane. Dan or other trainers use Robo-Crane to drop mealworms behind, urging the young cranes to follow the trike. When the daily session is done, the crew travels to the other pen site. They start again with the other cohort.

COSMOS
Canon
OPERATION MIGRATION
canadensis
OPERATION MIGRATION

It is a slow process. The chicks are nervous when the wing is added to the trike. Soon they get used to it. At first, the cranes walk slowly behind the plane. Then they run, flapping their little wings. Then they hop. At last, they pump their wings and lift off. They are about 60 days old when they start to fly.

The young birds, called *colts*, get excited when they hear the trike coming. They are let out of their pen as the trike taxis by the pen. They open their wings and charge down the grass after it. Joe says, "We try to make a tight circle and land again before they run out of energy."

What if they get tired? They land in the marsh. But as soon as they can, they fly back to the plane on the grass runway.

Some mornings are too foggy for the planes to fly. The birds can't train then. But they must build their strength for the migration. On foggy days, they are let out of the pen to fly on their own.

Swamp Monster!

Sometimes a chick has its own plans. It may land in the wetland by the pen instead of landing with the trike. The trainers become Swamp Monster to stop this. Dan and helpers hide in the wetland, covered with a camouflage cloth. If the cranes head for the marsh, Swamp Monster rises up with great noise and fuss. The wayward birds quickly learn to follow the plane instead!

From Two Flocks to One

By the end of August, both cohorts fly well behind the trike. It is time to start putting the two groups together. On September 5, 2001, the pilots lead the five older birds by air to join the younger birds. Now all are at one site. They will still train separately until they can fly for the same length of time. From two pens, they slowly get used to one another.

The birds grow taller and get stronger. They stay in the air longer. Most importantly, they learn their surroundings from the air. This will help them know where to return in the spring. This is the place where they will someday mate and hatch chicks. And they will teach their own chicks their migration path. In time, these birds will help build the new eastern flock of migrating whoopers!

chapter three

Countdown to Migration

Migration Checklist

When will they leave? Much needs to be done. The two cohorts still need to work out their new **pecking order** because they must fly as one group. They must build up their flight time. They will also need health checks and leg bands before migration. The pilots dread this. Joe says, "The birds have learned to trust us. We feed them treats and take them flying. For banding, we must do the unthinkable. We must grab the birds and put hoods over their heads so they can't see. They will not be pleased with us."

Unforeseen Events

September 11, 2001, is set for health checks and banding. The crew works in silence. They must hold each bird still for 30 minutes or more. Blood is drawn. Wings and legs are checked. Each bird gets color-coded leg bands on one leg. The other leg is fitted with a band that gives off radio signals so the birds can be tracked.

Meanwhile, shocking events stun the world. Terrorists attack the United States. They crash hijacked planes into the World Trade Center in New York. They crash a third hijacked plane into the Pentagon in Washington, D.C. A fourth goes down in Pennsylvania before reaching its target. In the end, nearly 3000 lives are lost.

Far away at Necedah NWR, another death takes place. The stress of being handled is too much for Crane 11. By midnight the bird is dead—stressed to death. Everyone is sad. He was one of the lead birds in the flock.

Joe Duff is sad, but he says, "This project is, above all else, an experiment. We have to accept losses and try harder. We recognize that birds may be lost or killed during the journey to Florida. Every effort will be made to prevent losses. But the project is going ahead with the knowledge that some birds may not survive."

Setbacks

The 9/11 attacks on the United States affect everything. Aircraft are grounded for days after the 9/11 attacks. At Necedah, crane training screeches to a halt. Joe adjusts the planned route due to new air rules.

Other setbacks happen. The pilots and handlers must rebuild the birds' trust after the stress of banding. The birds' legs are sore. They now wear leg bands and radios. This makes them reluctant to fly. The pilots work hard to win back the cranes' trust. The weather is bad for over a week. Deke and Dan are at their distant home on visits. After 9/11, they are still at home, stranded there by the ban on air travel.

The cranes and planes fly again on September 22. It's a good day. They fly for eight minutes. Joe says, "Between now and the start of migration, we will fly as often as possible to exercise them. We will increase their endurance until the season changes and it is time to head south."

Now There Are Eight

More worries. Crane 4 is a rebel. He drops out at the same spot during each flight. Is it a bad habit? Or isn't he strong enough to migrate 1225 miles? This wayward bird is watched with concern.

Crane 9's left wing droops. She has trouble flying with the other birds. A health check shows a problem with her feathers. Would she be able to fly the distance? The team thinks not. So Crane 9 is sent on a plane to the New Orleans Audubon Zoo. Eight cranes are left.

Crane Festival

On September 29, the public gets its first glimpse of these historic cranes. The town of Necedah holds a Crane Festival. The crowd watches in silence as the pilots and cranes fly over in perfect formation. The project has been done in near secrecy to shield the birds from human sights and sounds. Still, public support is huge. Today's celebration lets citizens share the pride.

Good to Go?

The migration target date is set for October 15. By now the routine is smooth. Caretakers walk to the pen as the planes taxi or fly to the pen gate. The trikes' speakers play the contact call to say, "Follow me. Let's go!"

The doors to the pen are thrown open. The cranes race out in a blur of white feathers and black wingtips. They jump and flap their wings, ready to go. The pilots rev up their engines. Then they're airborne with eight trusting birds behind.

Stopover Sites

Stopover sites for the migration are set. The birds must have secure places to rest and **forage**. About 25 private, state, and federal lands will be used along the route.

By October 4, the tiny flock sleeps, eats, and flies together. Their flight time is up to 12 minutes. But Crane 4 often does not care to join them.

October 15 arrives. The birds' longest flight up to now lasted only 27 minutes. With headwinds blowing at 19 miles per hour, today's flight will take 90 minutes. The young birds cannot do this. The migration must wait.

October 16 brings choppy air. This would mean a rough flight for the birds. Again the pilots and trainers postpone leaving.

Flying Made Easy

In the wild, young cranes would follow their parents. The families of soaring birds would leave around noon. The sun's heat is strongest then. The heat makes rising swirls of air called thermals. Cranes can soar on these *thermals*. Birds riding thermals seldom flap their wings, so they can stay in the air for hours. They can cover a few hundred miles with little effort.

Three Trikes and the Team

The migration will use three ultralights. One is the lead plane. Joe Duff and Deke Clark will take turns leading the flock. A chase plane follows the lead plane. Flying chase, Deke or Joe will bring back cranes that get tired or break from the group. The third trike is the scout plane. Bill Lishman is the scout pilot. He will check the landing site and clear the way. No people, animals, or vehicles can be near the cranes. Their survival depends on fear of humans, because humans pose a great danger.

A ground crew will travel with the team's motor homes. They will bring the birds' night pen and other equipment. Heather Ray will keep the pilots and crew organized. Many volunteers will help too. The humans outnumber the birds two to one.

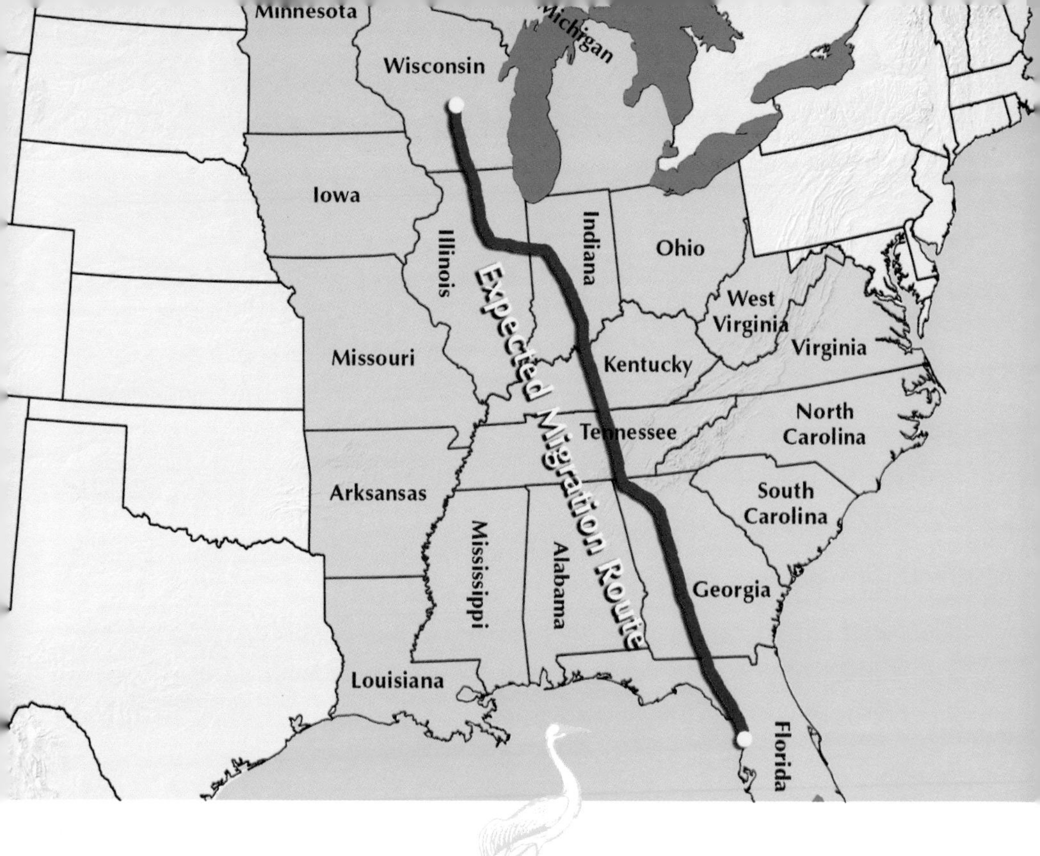

chapter four

Gone South! The Historic Migration

October 17, 2001: Day 1

We Have Liftoff!

At 7:15 a.m., eight whooping cranes and three trikes take off at last. The historic migration is under way! In the air, eight cranes line up behind the trike's wing. On the ground, the trailers and trucks race along the roadways to the first planned stopover site. The crew must set up the pen and get out of sight for the cranes' arrival.

The clear, cool weather is perfect for flying. Light winds from the northwest help push the tiny flock along. At 7:59 a.m., seven cranes land in Adams County, Wisconsin. They have traveled 29.3 miles south. Once again, Crane 4 shows his rebel streak. He breaks away from Deke's wing and drops out. Scout pilot Bill radios the bird's location to the ground team. Trackers pick up the signal from the radio band on the bird's leg. They find Crane 4 back near the Refuge. Was he trying to go back home? It's tricky work, but they capture Crane 4, place him in a crate, and take him to his flock.

The crew thinks Crane 4 is a bad example. What if he leads other birds astray? Trucking Crane 4 to the next stop may be wise.

Ground Crew

The ground crew packs up the overnight pen when the cranes are airborne. They get the campers and trailers set to go. The ground crew travels slower than the birds. That means the pen isn't up when the cranes land. The pilots, still in costume, must walk the birds to a spot where they can forage. The cranes must not see the ground crew at work.

The ground crew scrambles to set up the pen and feeders. They make it all look as wild and natural as possible. The pilots walk the birds to their night pen. Then they join the ground crew. Now everyone eats, cleans up, wraps up one day, and prepares for the next.

October 18, 2001: Day 2

Too Windy to Fly

No travel today. Gusty winds keep the crew and cranes on the ground.

October 19, 2001: Day 3

No Progress South

The crew decides not to fly Crane 4 today. Safe in a crate, he travels by truck. The trikes take off with seven cranes at 7:23 a.m. A strong headwind comes up. Joe and Deke climb higher in search of smoother air. They also want to avoid flying over a busy highway. The smoother air doesn't last long. Three birds soon turn back. The pilots think this is a good idea! The planes and cranes return to their old site in Adams County. Will weather be better tomorrow?

October 20, 2001: Day 4

Noisy Trucks Stop Cranes

The cranes and planes get off to a great start. They fly low in hopes of avoiding headwinds up higher. But the cranes hear the large, noisy trucks on Interstate 90/94 below. The roar of traffic scares the young birds. They have never heard that sound and scatter in every direction. But they follow Deke and Joe, as long as they don't try to cross the highway. It's too windy to fly higher. The pilots try again to fly over the highway. Each time the birds refuse to cross the noisy river of trucks. Twelve miles later the pilots give up. They turn back again to Adams County.

October 21, 2001: Day 5

Up and Over the Highway

Deke climbs to 400 feet and leads the cranes across the highway. Higher up, they find smooth air at last. Winds from the north help push them along. Seven birds fly 21.4 miles in 41 minutes. Crane 4 goes by truck. He joins the flock at the next stop—Sauk County, Wisconsin. The cranes have gone 50.7 miles of the total route.

October 22, 2001: Day 6

On to Stop Three

Hooray for tailwinds! The flock flies for 1 hour and 52 minutes. They add 48 miles to the journey. Total distance: 98.7 miles. In Green County, Wisconsin.

Flying Time

Why don't they fly longer on good days like this? Ultralight flights are limited to the calm air of early morning. These cranes learn to use the wake (like a boat's wake in water) off the wing of the trike. This can only happen when the air is smooth and the wing stays still. If the air is rough and the wing bounces around, the birds must move away and follow from a safe distance. They are forced to flap their wings. They soon get too tired or too hot.

October 23, 2001: Day 7

Grounded by Fog

No travel today. The cranes forage. The crew knows no such thing as a break. They take care of trikes, trucks, birds, and business.

October 24, 2001: Day 8

Still Down

Fogged in again, the cranes are content. The crew feels dismayed by delays.

Night comes. A fierce wind starts to blow. The wind rocks the motor homes where Deke, Dan, and Joe stay. Wind swoops into the valley where the bird pen stands. Deke and Dan go to check the birds. After an hour, Joe starts to worry. He leaves to check on his partners. The wind has blown the pen apart. Deke and Dan have found only three cranes in the wreck of the pen. Luckily, these birds are unhurt. They put the three cranes inside the sturdy pen trailer. Dan and Deke set out to search for the birds. The storm rages. Joe meets Dan and Deke on the road. They race back to camp and call for help to track the lost cranes. Bird specialist Kelly Maguire, project biologist Richard Urbanek, and bird doctor Julie Langenberg come. Everyone picks up costumes, vocalizers, dry boots, radios, tracking receivers, and night-vision scopes. Then they head back into the dark.

They hike through brambles and trees. They are wet, cold, and covered in burrs. As they walk, they play the brood call. Then they stop and listen for crane cries over the howling wind. By 2:30 a.m., they have all but one bird back in the patched pen. The pen shakes in the gusting wind.

The trackers get no signals on Crane 3. The bird's solar-assisted radio band does not work at night. Dan settles for the night in the trailer near the birds.

October 25, 2001: Day 9

From Bad to Worse

Joe and the others are back to search at dawn. Dan goes back to warm up at the base camp. Kelly and Julie drive nearby roads. Finally, a signal! They meet Joe on a hilltop. After an hour's search, Kelly finds the body of Crane 3. The bird lay dead under a power line that it hit in the dark.

The migration has been under way one week. Just under 100 miles have been gained. Troubles are many. Frost to scrape off the plane. Fog that dampens moods. Highways that the birds would not cross. Headwinds that the birds would not fight. Rain and lightning. Windstorms and power lines. Delays and death.

October 26, 2001: Day 10

Four Days Down

The crew spends another grim day in bad weather. They are still hunkered down at the Green County site. They know that the loss of a crane is always possible. This migration is an experiment. Yet it has a good chance for success. The migration will go on. Joe Duff writes in the logbook. "All we are missing is an earthquake, a tidal wave, and anything that remotely looks like luck."

October 27, 2001: Day 11

Crossing to Illinois

At last! Tailwinds make a good day to fly. Only the lead and the chase planes fly. Bill's plane is giving him trouble. Without Bill's plane to help if Crane 4 drops out again, the crew decides that the bird should go in the truck.

The cranes double the distance they've made so far! They soar 94.7 miles into LaSalle County, Illinois. Airtime is 1 hour and 55 minutes. Average speed is 59 miles per hour. That's a record!

Canadian/Texan Flock

The tiny flock's wild whooper cousins are also in the air. The first eight arrivals from the main flock in Canada have reached their winter home in Texas.

October 28, 2001: Day 12

Too Windy to Fly

The pilots fly a test flight with Crane 4 to see what he does. The crane flies with the trike—a good sign. Maybe he can still fly with his flock. But strong winds blow from the southwest. The birds would have to work very hard to make any gains in such winds. It's simply too windy to fly. They stay in LaSalle County.

October 29, 2001: Day 13

A Midday Flight

Winds blow from the south—the wrong direction to help the cranes. But their luck changes when the wind changes. They make a fast takeoff at 1:08 p.m. One hour and 46 minutes later, they land 61.6 miles away. This is the last stop in Illinois: Kankakee County. Crane 4 rides in the truck. But the crew will try him with the flock again when they fly over flatter land with fewer trees. That way, he's easier to track if he decides to go his own way.

October 30, 2001: Day 14

Headwinds Again

Flying today would be a waste of time and energy. The crew waits at stop 5 in Kankakee until the winds are favorable.

The crew creates a puddle for the cranes. The birds like to play with the hose. They eat from two feeders. This gives submissive chicks a chance to use the other feeder.

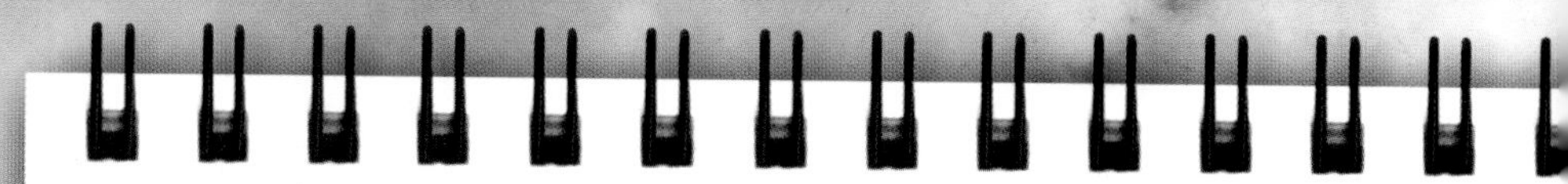

October 31, 2001: Day 15

Still Grounded

Today is rainy. Will tomorrow bring good flying weather? Everyone hopes! They are weary of delays.

November 1, 2001: Day 16

More Winds

Today's winds blow from south to north—wrong direction again! So far, noisy trucks held up the journey one day and weather held it up for ten days. Last year's test migration with sandhill cranes had only seven days off for poor weather.

November 2, 2001: Day 17

Still Down

Grounded again by fog and high winds. Kankakee, Illinois, is getting mighty familiar. They are 255 miles from Necedah.

November 3, 2001: Day 18

Indiana Bound

Airborne again! Crane 7 drops out early in the flight. Joe explains why. The birds are let out of the pen just seconds before takeoff. They sometimes get tangled in the rush. So Crane 7 is behind from the start. She can't catch up, so she drops out. The tracking crew finds her. Crane 7 joins Crane 4 and travels by truck.

The birds set a new record flight time of 2 hours and 9 minutes! They fly 91.4 miles. At 8:32 a.m., they land in Boone County, Indiana. That's 346.4 miles total.

November 4, 2001: Day 19

A Day Off

The cranes rest after yesterday's long, warm flight.

Heat and Whoopers

Warm weather makes flying hard. The birds pant when they get hot and tired. How else do the cranes show they are tired? They change flight order, looking for an easier place to fly. They spread their toes to help cool their body. They drop lower as they try to keep up. Then the chase pilot helps. He moves in so the tired bird gets a lift near his wing. A bird close to the wing has an easier flight. The plane creates some wind beneath the tired bird's wings.

November 5, 2001: Day 20

Another 57 Miles!

Frost must be scraped from the trikes' wings. At last, the ultralights lift off at 7:42 a.m. They climb to 1400 feet. Tailwinds help up here. When the wind changes, they drop lower. Headwinds slow them. They land at 9:50 a.m. in Morgan County, Indiana. This is site 7 in the 1225-mile journey.

November 6, 2001: Day 21

Good to Go! Two in a Row!

Frost on the wings causes a late take-off. But for the first time since October 22, the birds fly two days in a row. They log another 43 miles today in 1 hour and 40 minutes. Their stop is in Jennings County, Indiana.

November 7, 2001: Day 22

Great Day!

The sun shines and the air is still. Six birds fly an awesome 91.2 miles! The pilots even skip over a stop and head to the next one. Stop 9 is in Washington County, Kentucky. Today's flying time is 2 hours and 2 minutes. That's 537.6 miles gone. How many to go?

The cranes have shown they can fly about 100 miles a day. This is great news. It means they can make up for weather delays.

November 8, 2001: Day 23

Hello, Kentucky

Sunny and cool—a good day to fly! Airborne at 7:42 a.m., the cranes fly 1 hour and 37 minutes. Adair County, Kentucky, is the stop. They add 54.5 miles. The total is now 592.1 miles.

Crane 2 breaks off about 10 miles into the trip. The bird seems tired. Deke veers off after her, and she flies back near Deke's wing. Crane 4 still rides in a truck. He may get another chance to fly when the route is free of forests. They don't want to lose him.

Meet Top Cover Pilots, Paula and Don

The trikes fly low and slow to lead the cranes to Florida. But a low plane can't see other planes around. So pilots Paula and Don Lounsbury fly "top cover" in their fast Cessna. Each day, Paula and Don take off before the others even wake up. They check things out. Then they keep watch well above the trikes and cranes. They clear the pilots through air traffic control zones. They keep in touch with the ground crew by radio. All this helps keep the cranes safe.

November 9, 2001: Day 24

Troubles in Tennessee

Six airborne cranes and one in a truck leave Kentucky. It's a long hard climb to clear the ridge ahead. Crane 6 soon falls behind. He breaks away. Deke and Bill turn back to round him up. Alone, Joe leads the other five into Tennessee. It's tough going. The air is rough and warm. Still 70 miles to go. Without Deke in the chase plane, a bird that drops out will be on its own. The fields between ridges give way to forest. There's no place for Joe to land. No tailwinds come up, and they plod along slowly. Joe smells smoke and soon flies into a layer of smoke from a forest fire. It seems like hours until the stop site comes into view. Five cranes land in Cumberland County, Tennessee. They've flown 1 hour and 59 minutes.

Joe walks the birds to a pond. They probe in the muck for an hour. Now it is time to hide them. The ground crew will soon arrive. But another hour passes. Then Joe remembers that it takes the crew three hours to drive the distance he flew in two. At last, Joe sees the trucks. He walks the birds back to the hiding place. When the pen is up and the coast is clear, he walks the birds to their night pen.

Then Joe gets the bad news. Crane 6 is lost in the Kentucky hills. WCEP crane experts Sara Zimorski and Dr. Glenn Olsen are called to help search for him.

The other cranes add 75.3 miles. They've gone about 667.4 miles. Over halfway!

November 10, 2001: Day 25

The Search for Crane 6

The crew is grounded. They search by air and ground for Crane 6.

Paula's Cessna flies with a tracking antenna. Helpers search with the ground team. Finally, good news arrives. It comes from the landowner of the last stopover site. Back in Adair County, Kentucky, she looks out her window. She sees Crane 6 on the grass runway! She calls the team, but Crane 6 takes off before Dan and Sara arrive.

Radio signals tell Dan and Sara that the bird is about 5 miles south, probably near a river. They play the contact call when they're close to the river. Crane 6 responds! They see the crane flying overhead. But he can't land because there are too many trees. Dan tries to help. He runs to the nearest hilltop clearing while playing the crane call. Soon the lost crane lands beside Dan, crying a mournful peep. Heather dubs Dan the crane's "knight in muddy white costume."

November 11, 2001: Day 26

Bumpy Air and Fog: No Go

Thick fog finally clears. A test flight finds bumpy air. Tennessee is hilly and the cranes must cross a high ridge ahead. It would be hard to get them to climb high enough to cross the ridge in such rough air. They **stand down.**

November 12, 2001: Day 27

High Ridge Challenge

Soon after takeoff from Cumberland County, Tennessee, one bird breaks off and turns back to the site. Deke gives chase and lands with the bird. Deke crates the wayward crane, which joins Crane 4 in the truck. This time it's Crane 7!

Joe keeps flying with five cranes. He climbs higher with them. They need more altitude to cross the mountains. It is tricky, but he leads the birds over the ridge. They fly on through a valley. Another ridge is 25 miles farther. The warm air gets rougher as they go. The birds balk at climbing higher still. They won't do what it takes to get up and over. Joe sees that the birds won't go. He radios Bill. They look for a new landing spot in the vast forest valley below. The ground crew gets the news. They change their route too. All land safely in Bledsoe County, Tennessee. This is 26 miles south of the last site but 18 miles north of the planned stop. That's 693.4 miles gone in the 1225-mile journey.

November 13, 2001: Day 28

Over the Top!

The cranes conquer their toughest challenge. They make it over the 2800-foot mountain ridge that stopped them yesterday. The pilots lead them skillfully as they circle upward. They fly a very slowly—21 mph—as they climb over the ridge. But Crane 5 turns back. Scout pilot Bill Lishman leads the bird back to last night's pen. Crane 5 then joins Crane 4 on the road to stop 13 in Meigs County, Tennessee. Flying time is 1 hour and 20 minutes. They cover 17 miles for a total of 710.4 miles.

Too Cold, Too Hot

Joe and Deke fly in warm clothes today. At 34°F, it's cold. They wear snowmobile suits and hats. Gloves, vests and scarves. And always the white costume. But the air warms to 60°F by the time the cranes go in the pen. The pilots gladly get out of crane sight and peel off the layers. Everyone needs a rest.

November 14, 2001: Day 29

Georgia!

The cranes fly 1 hour and 45 minutes. They cross another state border. Now in Georgia, they add 67.2 miles. That means 777.6 miles gone and about 451 to go.

Today one bird drops out. Flying scout plane, Bill leads the bird the last 50 miles. Two birds fly on Deke's wing, and three birds fly on Joe's. All settle safely for the night in Gordon County, Georgia, stop site 14.

November 15, 2001: Day 30

Smooth Flying

Tailwinds today! The cranes cover 65.8 miles in 1 hour and 27 minutes. They arrive at 9:59 a.m. The total is 843.4 miles gone. Stop 15 is Coweta County, Georgia. Crane 4 still goes the easy way—in a truck. Maybe he's the smartest one!

November 16, 2001: Day 31

A Warm Day

The trikes and cranes leave with frost and fog. They fly 40.8 miles in perfect formation. The air warms to 60 degrees, so that's far enough. Warm air can overheat the hardworking birds. They stop at 9:42 a.m. at site 16 in Pike County, Georgia.

November 17, 2001: Day 32

Stop 17 on the 17th

Are these lucky numbers? The young cranes blow out of Pike County with cool air and great tailwinds. One hour and 28 minutes later they reach their next stop. They add 52.1 miles. Total: 936.3 miles.

November 18, 2001: Day 33

Last Stop in Georgia

No frost, no fog. The team gets an early start. They soon run into headwinds that slow them down. Still, they cover 45.7 miles. They fly 1 hour and 32 minutes. Joe lands with five cranes and Deke lands with his usual loner—Crane 6. They've come 982 miles. Terrell County, Georgia, is stop 18. The next stop should be in Florida!

November 19, 2001: Day 34

Unflyable Weather

All is ready for takeoff. Then top cover pilot Paula radios STOP. Checking conditions before the cranes fly, Paula finds warm air and headwinds. It would be a hard flight for the birds. Today's plans for a 60-mile flight will wait. The team hopes tomorrow is better.

Crane 4 will finish the trip to Florida in a truck. The team can't risk the crane dropping out again. He may tempt another bird to go with him. But he will spend the winter with his flock. The team hopes Crane 4 will fly with the flock when they migrate north in spring. He flew well with the flock when they took flight on their own. The team wants him to stay in the experiment.

November 20, 2001: Day 35

Grounded Again

It will be too hot to fly by the time the heavy fog is gone. This is day 17 of zero miles gained. Will the flock cross into Florida tomorrow?

November 21, 2001: Day 36

Passing 1000 Miles!

Helped by tailwinds, six cranes fly for 1 hour and 34 minutes. Not bad in the warm Georgia air. Today's flight puts the total at 1042.1 miles! Stop 19 is Cook County, Georgia. Florida is 60.1 miles closer! When will the cranes' journey end? The crew now guesses Monday, November 26.

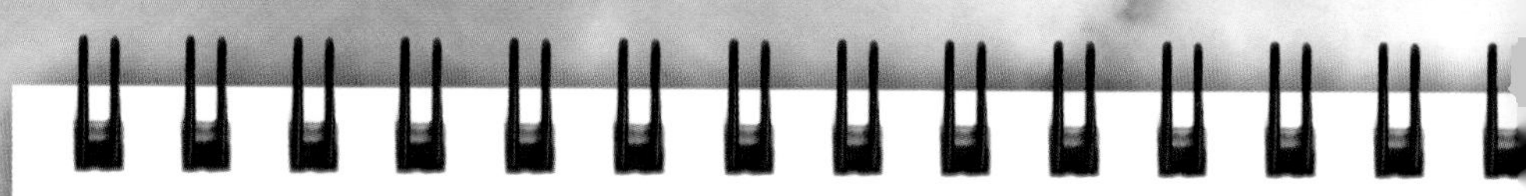

November 22, 2001: Day 37

Another Day on the Ground

Joe takes a test flight after the fog lifts. He figures a ground speed of 16 mph. This means a 2.5-hour flight to their next site. That would push the young cranes to their limit. So the crew spends this Thanksgiving Day in Cook County, Georgia. Florida is still 60.1 miles away.

November 23, 2001: Day 38

Rained Out

Rain stops the migration today. At site 19 in Cook County, Heather sums up the trip so far. "Of the 38 days, we have made progress on 19 and stood down for an equal number—16 for weather, 1 due to noisy trucks spooking the birds, 1 to rest, and 1 to find Crane 6 in Kentucky. We've covered 1042.1 miles. The pilots and cranes have spent 29 hours and 38 minutes in flight. We have about 183 miles until our final stop. Compare this with last year's sandhill crane migration. That took 40 days. The pilots and birds had 31 days of flight. They had 9 down days—7 for weather and 2 for repairs."

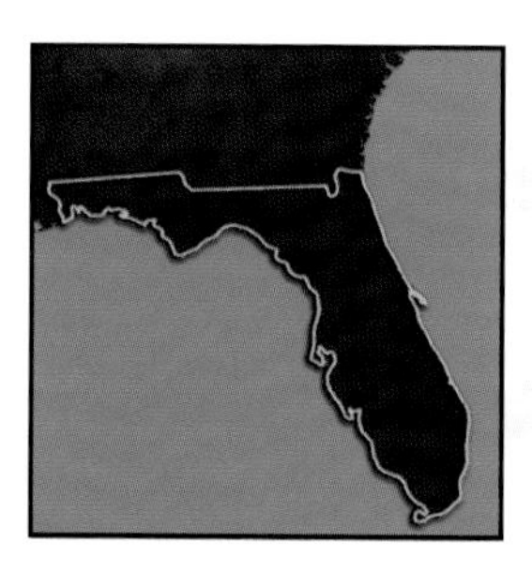

November 24, 2001: Day 39

Florida At Last!

The pilots and cranes finally cross into Florida. It's not easy! A wall of fog comes up. They land in a harvested cotton field and wait for the fog to lift. It's more than an hour. As the fog clears, the farmer comes to see what's going on. Luckily, they are airborne before the cranes see the human. Ten miles and 19 minutes later, they reach site 20. It's Hamilton County, Florida. They cover 38.6 miles. Total distance is 1080.7. Getting closer!

November 25, 2001: Day 40

Another Day Down

Today is day 17 of weather delays. Strong winds mean they'll make no progress and just get tired. The crew and cranes stay where they are. Last year, day 40 marked the end of the sandhill cranes' test migration. The young whoopers still have 144 miles to go.

November 26, 2001: Day 41

Airborne Again

They start late due to fog but add 20.9 miles. It takes 46 minutes. They fly low where there's less wind. People can see the planes and cranes. Joe's logbook says, "Three birds surfed on each wing as we passed over private homes on a Saturday morning. It is a common sight to see a homeowner standing in the backyard sipping coffee as we approach. We see him drop his coffee and dash for the back door. As we pass the house, the family in pajamas charges out the door with cameras in hand."

Stop 21 is in Suwannee County, Florida. They've come 1101.6 miles.

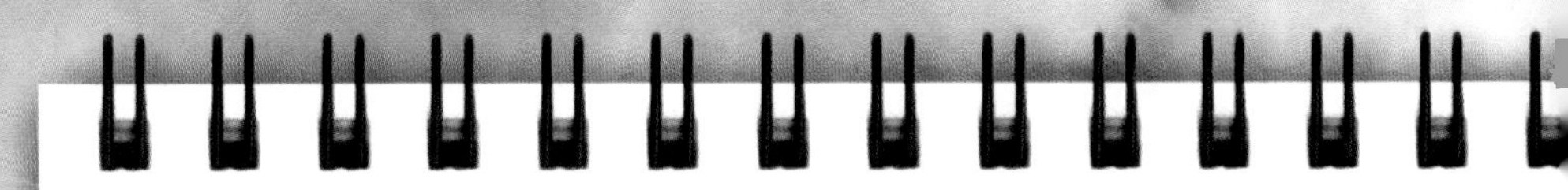

November 27, 2001: Day 42

A Few More Miles

Flapping their wings is exercise for the birds. It's 66°F, and they could easily overheat. Still they fly 45 minutes and gain 20.4 miles. When will they reach their winter home? This is Tuesday. The crew thinks they won't reach there before Saturday. They're at stop 22 and still in Suwannee County.

November 28, 2001: Day 43

Headwinds and Heat Turn Them Back

Today's heat and headwinds of 27 mph are stressful for the birds. The team flies only five minutes. They also note that Crane 5 holds his neck in a strange way. They turn back to stop 22. Another weather pattern is coming. They may not reach Chassahowitzka (chaz uh how ITZ kuh) National Wildlife Refuge this weekend after all.

November 29, 2001: Day 44

Two Cranes Trucked

Worried about Crane 5, the team crates him to travel by road. Five cranes take off in light fog at 7:17 a.m. They fly 20 miles. Now they've come 1142 miles. Stopover site 23 is in Gilchrist County, Florida. The doctor checks Crane 5. A neck X ray is normal. The bird's blood will be tested. They watch Crane 5 with care.

Injury Report

Weeks later, researchers decide that Crane 5 injured his neck by getting his long beak stuck in a leg band.

November 30, 2001: Day 45

No Flight Today

It's too windy to fly, but the weather looks good for the weekend. They stay at stop 23 again. Today is Friday. They hope Monday will be arrival day!

The team has been on the road for 45 days. Some have been away from families all summer to train the chicks. They are eager to see the end of the road! Dan writes: "I miss my wife, my dog, my cat, my parrot, my fish, and the house I left early in July—unfinished repairs and all."

December 1, 2001: Day 46

Stalled by Weather

The trikes are in the air. The handlers are set to let the cranes out to follow. The road crew is about to unhook the trailers and drive. Then Joe's radio brings bad news. Clouds, fog, and drizzle are now on the airstrip where the pilots had just taken off. "Roger that; stand down," say Joe and Deke. They turn back for another day going nowhere.

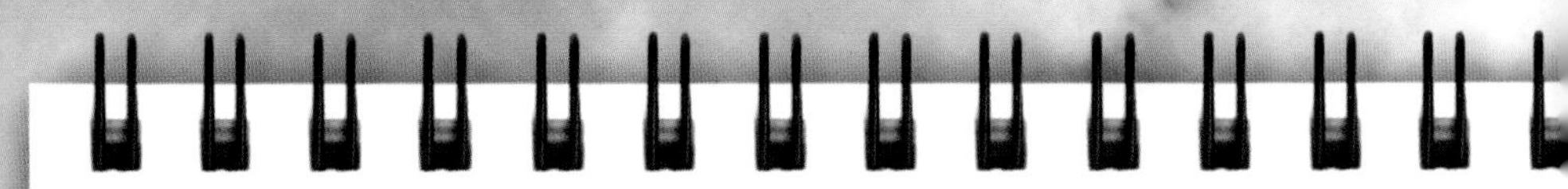

December 2, 2001: Day 47

Back in the Air

Good news at last! They take off with cooler temps and a slight tailwind. They land for a short time to wait out some fog. But then they're airborne again. They finish the trip to Levy County, Florida. Today's flight is 55.4 miles. Just one more flight to the finish line!

December 3, 2001: Day 48

A Day to Whoop It Up!

The young cranes fly the final 25.1 miles. The migration is officially over! About 1000 people come to a mall to see the cranes fly over. They wait in silence, all eyes on the sky. Here they come! Six magnificent white birds following a funny yellow plane. After they pass, Heather gives the nod. The crowd hoots and claps and cheers. Many shed tears of joy at the thrilling sight.

The young cranes are 48 days and 1222.5 miles from their start. All seven go to a temporary pen for the night. In a day or two, Deke and Joe will fly them to their final landing site and new home.

December 4, 2001

One Crane Home

Who would have guessed that Crane 4 would beat all the other cranes by a day? He is first to travel to the cranes' new home. Tucked in his crate, he is loaded onto a boat. He cries loudly over the engine noise. The boat heads out to a remote island. Dan and two other experts go with Crane 4. At the island, they carry his crate through the rushes and black muck. The permanent pen is here. Crane 4 is released in the large pen with the open top. Then they use the crane puppet to lead Crane 4 to the part of the pen with a top net. The happy bird probes in the muck. He finds tasty snails. He spends this night alone.

December 5, 2001

MISSION ACCOMPLISHED

It's 7:30 a.m. The cranes wait eagerly for Joe's trike to pass low by their pen. On the mark, get set, fly! They climb higher with each stroke of their long, elegant wings. They catch up to the tiny plane. It is their trusty guide. Soon the birds glide over the vast wetlands of Chassahowitzka National Wildlife Refuge. Dan is already there. He waits at the pen to "call them in."

Joe makes a low pass over the pen. Halfway down, he turns off the "follow me" call. Dan's vocalizer calls the young cranes down—and HOME. Joe climbs steeply. He veers off as the birds start their landing. All six land in or near the pen.

Once inside, they stay in a top-netted part. They get a final health check. Satellite transmitters are put on three cranes. These can help track the birds' movements over the winter and spring. Then the top net is removed. The birds are free to fly and explore.

Heather sends out the news. "The 'Chass-seven' have arrived. And with them, hope for survival of the species."

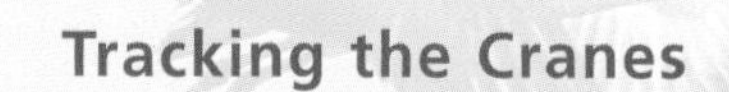

Tracking the Cranes

All the birds wear leg bands for radio tracking. Three of the birds—Cranes 2, 4, and 5—also get satellite-tracking devices called *PTTs*. Satellite data can tell trackers where to look for a missing bird using radio signals if they don't already know where the bird is. The Team wants to keep tabs on what habitat the birds use. But the main goal is to find out what route they take back.

Why did Cranes 2, 4, and 5 get the satellite transmitters? It was a careful choice. Cranes 2 and 5 were leaders and strong flyers. The team hoped these birds would lead on the way back north. The other birds will likely follow these two. Crane 4 may take his own path in the spring. The PTT will help WCEP know where he goes.

chapter five

Winter in the Wild

On Their Own

The flock uses a permanent pen at their island refuge. No one wants these valuable birds to be a predator's meal. The pen has an open top. A high fence surrounds it. About two feet of fence is buried under the ground. This stops alligators and other predators from digging under the fence. An electric wire adds more protection.

The pen has a feeding station and a big pond. A costumed dummy stands in the water. The cranes know the costume. It's there to encourage them to **roost** in the pond at night. Roosting in water is safest. The birds can hear the splash of a predator coming at them. They can fly off to escape.

Keeping Watch

A two-person team watches over the cranes. They are project biologist Dr. Richard Urbanek and Anne Lacey from ICF. Richard and Anne check on the birds every evening. An airboat brings them to the cranes' remote island. They check the pen and the food at the feeding stations. The two wear the white costumes whenever the birds can see them. Anne normally stays in a hidden place and runs video equipment to watch the cranes. She says, "We're trying to minimize contact, even with the costume. We don't want them to rely on it or think it will lead them back."

Learning to be Wild

What do the cranes do all day? They learn how to eat crabs, for one thing. A whooper's most important food is blue crabs. Parents must teach the young cranes how to eat the crunchy critters. So Anne or Richard catches crabs in a net. They bring the crabs into the pen. Then the crane puppet shows the young birds how to peck the crabs apart.

Richard and Anne keep careful notes. One such notation reads, "On December 12, Richard caught a small crab. He offered it to the birds. They were all interested, but Crane 4 grabbed the crab and gobbled it up." Anne and Richard later see the cranes catch little crabs. They have mastered the art of crabbing!

Watching Over the Birds

Anne Lacey rotated with Marianne Wellington, Sara Zimorski, and Kelly Maguire during the 126 days spent in Florida.

Bobcats and Bad News

Refuge staff work hard to keep the seven birds safe. They

set live traps for bobcats when tracks appear around the pens. Then come some shocks. A bobcat kills Crane 4 on December 17. Crane 10 is last seen alive on January 10, 2002. Her radio band quit working, but her remains were found on January 22. She was the victim of a bobcat. The "Chass-Seven" has become the Florida Five.

The rest of the winter passes safely. Spring comes. Many birds begin to migrate overhead. But not the young whoopers.

Richard and Anne are ready to take off when the cranes do. They will track the cranes as they fly north all on their own. Each of them will drive a truck with radio tracking equipment.

Everyone wonders. When will the young cranes head north? WILL they head north? "It's a guessing game," says Anne. "They are being very good birds, staying around their pen area, foraging in the tidal pond in their pen, and catching crabs."

chapter six

Wisconsin, Here They Come!

They're OFF!

The Florida Five make their move on April 9 after 126 days in their winter wetlands.

Anne checks their position by radio signal when she notices they are moving. With good tailwinds, the birds move FAST. Jumping into separate trucks, Anne and Richard begin the chase. The migration is under way!

"The birds were really active last night. Perhaps that was their way of saying they were ready to go," said Anne. "They left the pen several times. We also heard them calling one another much more often than in previous weeks."

Will they find their way back to Necedah NWR? How long will it take? Will all five survive?

Knowing When to Go

How do the cranes decide when to go? They have inner clocks. They have an ability to know how long the days are. It's a mystery how animals can measure the length of daylight. But people have inner clocks too! How else can they explain that they wake up or get hungry about the same time each day?

The Journey North

April 9: The pioneer flock flies almost 7 hours on their first day. The five whoppers settle in Wilcox County, Georgia. They are about 217 miles from their Florida feeding grounds. They choose a good place to forage and roost.

April 10 and 11: They stay put and wait for better flying weather.

April 12: The birds fly from their roost area back to forage in their April 9 landing spot. Flying conditions are not the best. But the cranes seem determined to migrate. After nearly five hours of flying, they land in Henry County, Georgia. They cover 108 miles more.

April 13: They forage and wait for better flying weather.

Tracking the Return

Anne and Richard travel in two trucks. They planned this in case the birds separate. All five cranes still have someone tracking them.

April 14: The cranes cross the Tennessee state line at 4 p.m.! Just before that, Crane 7 peels off on her own. She flies solo and lands in McMinn County, Tennessee. She flies 144 miles in 6.8 hours. The group of four lands in Fentress County. They cover 200 miles in a record 8 hours in the air.

April 15: The flock of four is airborne again. Conditions seem ideal. Richard follows their radio signals. They have another record day of 238 miles in 8.5 hours! They roost in shallow water in a pasture in Johnson County, Indiana. Anne follows Crane 7's signals still in Tennessee.

April 16: Despite the windy conditions, the group leaves Indiana. Later in the day, they reach Lake Michigan. The Fabulous Four circle for two hours, apparently trying to decide which way to go around the large body of water. Richard knows if they go east, they will end up in Michigan, on the wrong side of the lake. Finally, the cranes fly west over Chicago, Illinois. But where could they land? Richard later tracks them to a forest preserve pond. Today they fly 214 miles in 8 hours. Anne tracks Crane 7 to her roosting place in White County, Indiana.

April 17: The four stay in a wetland within Chicago. Crane 7 is in Jasper County, Indiana. "She's found a nice little pond," says Anne. "She is foraging away and resting up, I hope, to continue the journey north tomorrow."

April 18: All five birds end the day in Wisconsin! The group of four flies 123 miles in 5 hours today. Crane 7 makes her way alone. She's near the border of Wisconsin and Illinois.

April 19: Touchdown in home territory! Despite poor weather today, the four seem determined to get home. Flying 1175 miles in 11 days, four of the five whoopers complete their spring migration. Led by Crane 1, they soar into Necedah NWR at 6:37 p.m. They land near the pen where they were raised. Meanwhile, Crane 7 shares a wetland with sandhill cranes elsewhere in Wisconsin. She's taking her time to get home. Awe, joy, and relief reign as word spreads among the WCEP team and the public.

May 3: Crane 7, the solo female, arrives at her fledging grounds at Necedah.

Record Return Trip

The birds shaved miles off their return trip. Trackers estimate they flew about 1175 miles. How can this be? The cranes didn't have to obey air traffic rules set after the 9/11 attacks. They did not have to avoid certain airspace like the pilots did. They could fly a more direct route.

The Fabulous Five: Flown Away Home

It's the end of a historic and successful migration. Without any help, the whoopers will leave each fall and come back each spring for the rest of their lives. The cranes will mature at four to five years of age. Then they'll lay eggs and raise chicks to help the flock grow. They'll teach their chicks the route they learned from their funny-looking parents in a tiny plane and white costumes.

Riding the Winds

The young cranes had good natural instincts. They knew how to take rising currents of warm air called *thermals* to aid their flying. Riding thermals, the cranes covered about 200 miles each day! Joe Duff reminds us, "Whooping cranes in the wild are soaring birds and will go as long as the weather carries them—as long as the thermals last and they're getting a free ride. The thermals stop when the sun goes down, and so do the cranes."

Crane	Sex	Facts & Updates
1	Male	Most dominant, or top, bird. Often led the birds right behind the wing of the aircraft. Was also seen at the end of the line. Led the way north on the first migration in spring. Hung around the training site where the new 2002 ultralight chicks were learning to follow the trike. Paired up with Crane 2 and migrated south with her in fall 2002. The two even joined the new 2002 ultralight chicks for a five-mile flight behind the ultralight on November 24 in Tennessee. They spent their second winter (2002–03) together in Pasco County, Florida.
2	Female	Often aggressive and challenged the costume. Injured her beak so top and bottom do not meet at the tip. This made her recognizable in the air. Often in the lead position. Upon return to Necedah, tried to claim the new chicks' training site as her territory. Came back many times after being driven away by OM pilots. Teamed with Crane 1 for the summer, the fall migration, and their second winter together in Florida.
3	Male	Killed at stop 3 on the journey south. Flew into a power line in the dark after the pen was knocked down by high winds.
4	Male	Smallest of flock. Had a bad habit of dropping out of the group while

Crane	Sex	Facts & Updates
		flying. Dropped out on the first day of migration and encouraged Crane 6 to leave as well. Not trusted to fly with other birds on the rest of the migration. Traveled in a crane crate to each site but stayed with other birds in the pen. Killed by a bobcat in Florida in December 2001.
5	Male	Second most dominant bird in the flock. Trainers called him the policeman. He was the first bird to check out a newcomer to the pen. Often aggressive to caretakers who wore different shoes or boots. Migrated to Florida in fall 2002 and spent the winter in the pen with the new chicks that came in the second year of ultralight migrations.
6	Male	Dropped out of the flight on the first day with Crane 4 but managed to complete the flight after much effort from Deke and Bill. Dropped out again between sites 10 and 11 and was retrieved. Often flew alone off Deke's wing. Upon return, spent much of the summer by himself in a remote area of Necedah NWR. Migrated south in fall 2002 to Hiwassee refuge in Tennessee and didn't continue to Florida until January 2003. Then joined sandhill cranes and crane 7 in Madison County, Florida, before moving a few miles from them.

Crane	Sex	Facts & Updates
7	Female	Low in the flock's pecking order. From Tennessee, flew the rest of the journey north by herself. Came back to Necedah, but moved. Spent summer 2002 at Horicon NWR about 75 miles from there. First to return to Florida during the fall 2002 migration, arriving in her old pen on November 22. Moved and lived with sandhill cranes in Madison County, Florida, for the winter of 2002–03.
8	Female	Died before the birds shipped to Necedah.
9	Female	Injured a wing during early training at Patuxent WRC. Her feathers did not grow well enough to make the fall migration. Now lives at the Audubon Zoo in New Orleans.
10	Female	Most **subservient** bird in the group. Flew in both the lead position and last position during flight. Killed by a bobcat in Florida in January 2002.
11	Male	Dominant. Good at flying with the trike. Died of capture myopathy, stress from being handled, during the health check and banding before migration south.

chapter seven

New Hope for Whooping Cranes

What's Next?

WCEP hopes to teach the same route to a new generation of whooper chicks each fall. Ultralights will lead a new generation of captive-bred cranes south each fall until perhaps 2005. If the eastern flock grows as WCEP hopes, the trikes won't be needed after that. The flock will have enough **veteran** cranes to lead the migration. The older whooping cranes can be guide birds. They can show future released chicks the way. The goal is to build a flock of 125 birds by 2020, with least 25 breeding pairs.

It will be a giant leap for conservation when whooping cranes soar again through eastern skies. Imagine. In your lifetime, you may see "whoop dreams" come true!

You Can Help

This exciting story will unfold in your lifetime. Share it with others! These Web sites will keep you up to date:

Whooping Crane Eastern Partnership (WCEP)
http://www.bringbackthecranes.org/

Operation Migration
www.operationmigration.org

Journey North
www.learner.org/jnorth

Meanwhile, write letters to your representatives in Congress. Ask them to pass laws that support programs to save and protect endangered species and their habitats.

Share money. Many groups and WCEP members raise funds to help the whooping crane reintroduction.

Lend a hand. Do you live near Baraboo, Wisconsin, or Laurel, Maryland? You could volunteer at these whooping crane breeding centers.

International Crane Foundation
http://www.savingcranes.org/

U.S.G.S. Patuxent Wildlife Research Center
http://www.pwrc.usgs.gov/

Migration 2001

Date	Down or Fly	Miles Flown	Total Miles
10/15	down/weather	0	0
10/16	down/weather	0	0
10/17	fly	29.3	29.3
10/18	down/weather	0	29.3
10/19	down/weather	0	29.3
10/20	fly/but turn back	0	29.3
10/21	fly	21.4	50.7
10/22	fly	48	98.7
10/23	down/weather	0	98.7
10/24	down/weather	0	98.7
10/25	down/weather	0	98.7
10/26	down/weather	0	98.7
10/27	fly	94.7	193.4
10/28	down/weather	0	193.4
10/29	fly	61.6	255
10/30	down/weather	0	255
10/31	down/weather	0	255
11/1	down/weather	0	255
11/2	down/weather	0	255
11/3	fly	91.4	346.4
11/4	down/rest	0	346.4
11/5	fly	57	403.4
11/6	fly	43	446.4
11/7	fly	91.2	537.6
11/8	fly	54.5	592.1
11/9	fly	75.3	667.4
11/10	down/search	0	667.4
11/11	down/weather	0	667.4
11/12	fly	22	689.4
11/13	fly	17	706.4
11/14	fly	67.2	773.6
11/15	fly	65.8	839.4
11/16	fly	40.8	880.2
11/17	fly	52.1	932.3
11/18	fly	45.7	978
11/19	down/weather	0	978
11/20	down/weather	0	978
11/21	fly	60.1	1038.1
11/22	down/weather	0	1038.1
11/23	down/weather	0	1038.1
11/24	fly	38.6	1076.7
11/25	down/weather	0	1076.7
11/26	fly	20	1096.7
11/27	fly	20.4	1117.1
11/28	down/weather	0	1117.1
11/29	fly	20	1137.1
11/30	down/weather	0	1137.1
12/1	down/weather	0	1137.1
12/2	fly	55.4	1192.5
12/3	fly	25.1	1217.6

Glossary

captive-breeding referring to a program in which wild species are captured and then bred and raised in a special facility under the care of wildlife experts

captivity state of being held; not being free in the wild

contact call purring sound made by an adult crane to say, "It's okay; follow me."

endangered in danger of becoming extinct or ceasing to exist

eon length of time that is too long to measure

fledge to grow the wing and tail feathers necessary for flying; to fly on its own

forage to obtain food from a place by searching and rummaging

habitat natural conditions and environment, for example, forest, desert, or wetlands, in which a plant or animal finds what it needs to live—food, water, shelter, space

imprinting rapid learning process that takes place early in life that causes an animal to recognize and be attracted to its own kind

marsh area of low-lying, waterlogged land

mealworm larva of various beetles

migrate to move from one habitat or environment to another in response to seasonal changes and variations in food supply

pecking order social system in which some members of a group are established as superior to others

pioneer person or group that is the first to do something or create something new

predator animal that hunts, kills, and eats other animals

refuge region of protected habitat for wildlife

roost to rest or sleep

stand down to stop or to end

subservient in a position of lesser importance; willing to be a follower rather than a leader

veteran having experience or skill from doing something before

vocalizer small, handheld device that plays digital recordings

Index